ABOVE: *Haymaking on Lord Wantage's estate at Lockinge, Oxfordshire. 1906. A considerable labour force was required. men for building the stacks and women for raking the new hay together into cocks ready for loading.*
COVER. *Autumn sowing of grain. Detail of a print of 1818, from a drawing by J. M. W. Turner.*

AGRICULTURAL HAND TOOLS

Roy Brigden

Shire Publications Ltd

CONTENTS

Copyright © 1983 by Roy Brigden. First published 1983; reprinted 1985, 1987. Shire Album 100. ISBN 0 85263 630 X. All rights reserved. No part of this publication may be reproduced or transmitted in any form or by any means, electronic or mechanical, including photocopy, recording, or any information storage and retrieval system, without permission in writing from the publishers, Shire Publications Ltd, Cromwell House, Church Street, Princes Risborough, Aylesbury, Bucks HP17 9AJ, UK.

Set in 9 on 9 point Times roman and printed in Great Britain by C. I. Thomas & Sons (Haverfordwest) Ltd, Press Buildings, Merlins Bridge, Haverfordwest, Dyfed.

ACKNOWLEDGEMENTS
   All photographs are copyright of the Institute of Agricultural History and Museum of English Rural Life, University of Reading.

A hedger enjoying a meal break beside a pile of dead wood with the tools of his trade at his feet. Photographed in the period between the world wars.

The rakemaking yard and workshop of John Sims at Pamber End, Hampshire, in 1953.

INTRODUCTION

Although mechanised farming began in the nineteenth century, hand tools nevertheless maintained an important, if declining role. On the smaller farms in particular the economies of scale that justified investment in new machinery were less easily realised and so the older methods survived for longer. Even the large, capital-intensive holdings continued into the twentieth century to rely considerably on human muscle: only since the Second World War have mechanical handling techniques ousted a whole range of accessory tools, from the dung rake to the hay knife, while some operations, such as hedge laying, have never

been effectively mechanised. There was, therefore, not so much a dramatic transformation between hand labour and machine but rather a developing conjunction between old and new. The onset of two world wars and the rising cost of manpower tipped the balance progressively in favour of the new.

This book describes the broad range of farm tools and hand-operated machines preserved in museums and collections around Britain. Some aspects, notably livestock and dairying, have been omitted for they are covered by other Albums in the series.

ABOVE: *The force required to push a breast plough came from the top of the user's thighs, often protected as here by wooden boards, leaving the arms free to guide and support. With the long shaft angled in this manner, the blade lay flat and cut through the soil at a depth of about 2 inches (50 mm). From 'The Book of the Farm' by Henry Stephens, fourth edition, 1891.*
BELOW LEFT: *Wrought iron breast plough share from County Durham with riveted plate repair to fin. 16½ inches (420 mm) long.*
BELOW RIGHT: *Pads worn to protect the thighs when using a breast plough. They are composed of two slightly curving pieces of ash wood joined by leather straps and with a loop at the top through which passes the operator's belt. Each pad is 1 foot (305 mm) long.*

4

Horseshoe-shaped drainage tile from Berkshire. From 1826 to 1850 drainage tiles were exempt from a general tax on tiles, hence the word 'drain' stamped on the top

FIELD MAINTENANCE AND PREPARATION

The removal of a thin top layer of soil in order to burn the vegetative matter it contained was extensively practised up to the beginning of the twentieth century as a means of preparing wasteland and old pasture for cultivation. When the ashes were spread back over the land, the process could also help maintain soil fertility. In some areas, such as the Fenlands of the eastern counties and the Cotswold hills, enrichment of the soil in this manner was an important feature of regular cultivation management until the widespread introduction of artificial fertilisers.

By paring and burning, rough growth was more quickly and completely destroyed than if it were ploughed in conventionally and left to rot. For this reason, the method was most often used as part of a rotation for breaking up grassland in preparation for a crop of wheat.

Although horse-drawn paring ploughs were in use in the middle of the nineteenth century, the manual and more costly alternative, the breast plough, continued to find favour because of the greater thoroughness with which the work was accomplished. With the breast plough, known in Scotland as a *flauchter-spade*, the turf was raised in strips and turned up, every 18 inches (0.5 m) or so, by a twist of the plough handle, and was left there to dry. Small heaps were then collected for burning and the ashes either distributed over the land at once or else carted away to be spread at a later date with manure from the farmyard. Paring was normally carried out in early spring and as many as eight or ten men might be employed in a single field with an output per man of approximately ¼ acre (0.1 ha) per day.

DRAINING
Thorough under drainage was a prominent feature of nineteenth-century land improvement, helped from the 1840s by government grants, through the Public Money Drainage Act of 1846, and the invention of a practical machine for the manufacture of drainage pipes.

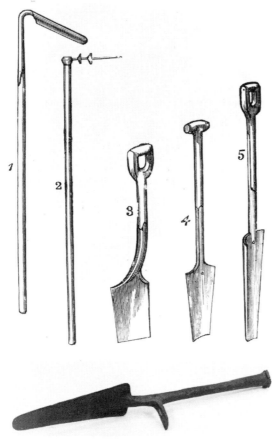

ABOVE: *A pull scoop for clearing mud and sludge from the bottom of drainage trenches. A wooden handle fitted into the tapering socket and was secured by two rivets. 14½ inches (370 mm) long.* UPPER LEFT: *Draining tools illustrated in J. P. Sheldon's 'Dairy Farming' of the 1880s. 1 and 2 are a pull scoop and a pipe layer. 3 to 5 are the three standard spades, with blades of diminishing width, required to dig a V-shaped trench.* LOWER LEFT: *Narrow drainage or bottoming spade of wrought iron with a socket for a wooden handle. A separate foot rest, for greater purchase, is provided at right angles further up the shaft because the shoulder of the blade is so narrow. The blade tapers to 1⅜ inches (35 mm) across.*

A common method of draining a field was to cut one main drain, 4 feet 6 inches (1.35 m) deep, along its lower edge. Water flowed into this from feeder drains traversing the field at a depth of 4 feet (1220 mm). The distance between feeder drains varied according to the soil conditions but 15 feet (4.5 metres) was the minimum. Both main and feeder drains were formed by digging trenches to the required depth, laying porous pipe tiles end to end along their length and replacing the soil.

V-shaped trenches were the most economical of labour and were dug with the aid of spades with long tapering blades. Three or more, diminishing in size, were needed for as the depth of the trench increased so its width was reduced. Hand or foot picks were used if necessary to loosen the soil before it was dug out, while the narrow seat for the pipe at the base of the trench was cut with a bottoming spade. The trench bottom was cleared and squared off with a long-handled scoop that could be operated from ground level. There were two types, both with similarly shaped blades, but the pull scoop was best for removing running mud and sludge, whereas the push scoop was preferable for drier earth and stones. A pipe layer, again fitted with a long handle, allowed the drainage pipes to be positioned from the ground alongside the trench. When all the pipes had been arranged, the trench was backfilled.

An iron plate, with leather supports, which was slipped over a boot and secured by a buckle to protect the sole during heavy digging from the discomfort that arises from continually bearing down on the shoulder of a spade. 8½ inches (215 mm) long.

HEDGING

Planting and maintaining field hedges, to provide stock-proof barriers, was a skill practised either by appointed members of a farm labour force or by specialist itinerant craftsmen. Joseph Arch, founder of the Agricultural Labourers Union in 1872, was a hedger who regularly found work in Gloucestershire, Herefordshire and Wales, as well as his native Warwickshire. Commanding a wage of at least 2s 6d (12½p) per day, he was better paid than the average agricultural labourer.

The hawthorn plant was much favoured for hedges because of its rapid growth, long life, hardiness and prickly stems that deterred the over-inquisitive

A 'Devon' type billhook with tanged blade and caulked handle for additional hand support. Made by Morris of Dunsford, Devon, in the early twentieth century. 14¾ inches (375 mm) long.

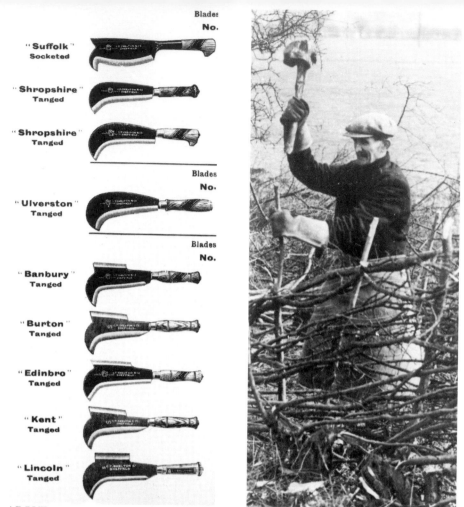

	Blades No.
"Suffolk" Socketed	
"Shropshire" Tanged	
"Shropshire" Tanged	

	Blades No.
"Ulverston" Tanged	

	Blades No.
"Banbury" Tanged	
"Burton" Tanged	
"Edinbro" Tanged	
"Kent" Tanged	
"Lincoln" Tanged	

ABOVE LEFT AND BELOW: *Billhooks (above) and two slashers (below) from the 1915 catalogue of the Sheffield tool specialists, G. T. Skelton and Company. A wide range of choice was common from the later nineteenth century as the large manufacturers achieved national markets by catering for all requirements and prejudices.*

ABOVE RIGHT: *A hedger using a maul to drive in a stake, cut from the dead wood, that acts as a support for the woven or plashed stems. Visible near the ground are the cuts made in the side of the stems to allow them to be laid over on the slant and create a solid fence line. Photographed in the south of England in the 1930s.*

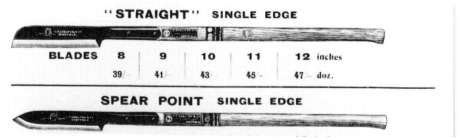

"STRAIGHT" SINGLE EDGE

BLADES	8	9	10	11	12 inches
	39/-	41/-	43/-	45/-	47/- doz.

SPEAR POINT SINGLE EDGE

A hedger's protective leather gloves. From Radnorshire (now part of Powys). Each one is 10½ inches (267 mm) long.

beast. But it required careful supervision for, if neglected, too much growth would go into height, where it was not needed, at the expense of thickness lower down. When tended correctly, a thorn hedge can function efficiently for a great many years.

Thorn plants two years old were used for new hedges and were set 6-12 inches (150-300 mm) apart in a trench that had first been cleaned with the pointed hedging spade. The young hedge required little further attention in its early years apart from the pruning of protruding stems along the sides and top to encourage it to grow thicker rather than larger. A light slasher, or switching bill, was used one-handed with upward strokes for this operation.

To lay a hedge was to restrain its vertical and outward growth in order to ensure that it was impenetrable nearer the ground. The task was carried out in the winter and early spring, and the best practice was for one twelfth of a farm's hedges to be laid every year so that each hedge received essential maintenance once every twelve years.

In detail, the practice of laying varied in different parts of Britain but, as an example, the rough and overhanging branches were first cut away with slasher and billhook to give access to the thicker main stems. Then every 2 feet (600 mm) a straight stem was chosen as a live stake and its top was cut off 4 feet (1.2 metres) from the ground. Where there was no suitable stem, a stake was cut to size from the dead wood already removed and driven into the required position with a maul. The first remaining stem not acting as a live stake was then bent over and an upward cut was made with a billhook near the ground on the side away from the bend. This prevented the stem from springing back while allowing it, provided the cut was not too deep, to remain in growth. Its length was then woven, on the slant, in and out of the upright stakes. Each stem in turn was plashed in among the stakes in this way to create a narrow fence that would fill out in subsequent years to leave a mature hedge free of gaps or holes.

9

LEFT: *A sowing sheet as commonly used in Scotland during the nineteenth century. The seed is held in the cradle of the left arm and the loose end of the sheet wound tightly around the left arm. The forward swing of the right arm to scatter seed occurs as the right leg advances. From 'The Book of the Farm' by Henry Stephens, fourth edition, 1891.*

BELOW: *A seedlip with sides made from a single length of deal bent to shape. The contour of the strap side allowed the container to be held close against the left side of the body while the vertical handle was gripped by the left hand. This one was used at East Hendred, Oxfordshire, until the 1930s. 25¼ inches (640 mm) long.*

BOTTOM: *Broadcasting seed from both a wooden (left) and a galvanised seedlip at Childrey, Oxfordshire, in the 1930s. Both men are at the same point in the rhythm of sowing, with the right hand gathering more seed from the container as the left leg advances.*

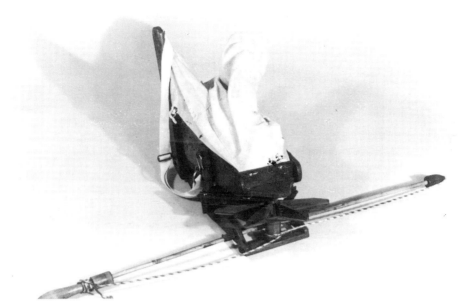

Twentieth-century form of the fiddle broadcaster. Clearly visible between the canvas seed container and the bamboo bow is the finned disc that scatters the seed. Overall length 20 inches (506 mm).

SEEDING AND WEEDING

Hand-operated devices for sowing seed took three main forms, those for broadcasting, drilling and dibbling, and all were in use during the nineteenth century.

BROADCAST SOWING

Scattering seed broadcast by hand is the simplest method of sowing although considerable experience and care are required for it to be performed efficiently. In Scotland a linen sheet running over the left shoulder and beneath the right arm was commonly used to hold seed in the cradle of the sower's left arm. But in England sowing baskets or seedlips made of basketwork, thin deal or, latterly, galvanised iron, were more usual.

Successful broadcasting depended upon settling into an even rhythm through the synchronising of arm and leg movements. Taking seed from the box in the right hand, the arm was swung backward as the left leg moved forward. The seed was then scattered progressively with the return swing of the arm as the right leg advanced. An inexperienced sower's efforts were visible to all at harvest time: some parts of the field would be thicker in crop than others while elsewhere bare streaks gave evidence of incorrect timing between successive casts. Broadcast sowing was a hazardous operation in windy weather, especially for grass seed, which, lighter than grain, was not so easily directed to its intended position.

An American development of the second half of the nineteenth century, the fiddle broadcaster, made sowing quicker and easier to accomplish well. Many have survived for they were used for grass and clover up to the Second World War and it is only in recent years that a Scottish firm has ceased to make them. The name was derived from the method of operating a bamboo bow at the front of the hopper box which caused a finned disc to make alternating revolutions and scatter seed forwards in a wide arc. Successful use of the fiddle broadcaster again relied on settling into a regular rhythm between

walking pace and movement of the bow. The width of cast varied according to the type of seed but the ideal distribution area ranged from 16 to 23 feet (5 to 7 m) and 2 acres (0.8 ha) or more could be sown in an hour.

A broadcast barrow gave a more even coverage of seed with less wastage than was usual with hand sowing. It scattered seed on the surface of the ground, to be subsequently covered by a harrow, as distinct from the drill, which deposited seed in straight rows at a uniform depth decided by the operator. For sowing corn, the broadcast barrow commonly took the form of a three-wheeled horse-drawn machine with a seed box 16 feet (5 m) or more in length. But for grass and

LEFT: *Broadcasting with a fiddle drill at Killinghall, near Harrogate, North Yorkshire, in the 1930s.*
BELOW: *A broadcast barrow in action. Just visible is the shaft taking the drive from the wheel axle into the seed box, where it drove a line of circular brushes. On many barrows there was a gear wheel of different diameter at either end of the axle so that the wheel could be taken out and replaced the other way round to vary the rate of flow.*

RIGHT: *A small Warwickshire seed drill made in the 1880s. Seed passes from the hopper into the dispenser behind, where a rotating grooved cylinder regulates the downward flow into the chute. The cylinder is operated through bevel gearing from the axle of the land wheels. As the drive shaft pivots slightly, the seed-dropping mechanism can be put in or out of gear as required.*

BELOW: *More recent seed drills have much in common with their nineteenth-century counterparts. The drill illustrated is from the 1941 catalogue of S. L. Allen and Company of Philadelphia and was marketed under the name of 'Planet Junior'. It was capable of sowing a wide range of fruit and vegetables and, together with other cultivating implements offered under the 'Planet' system, was popular with market gardeners. Many were sold in Britain and many examples survive. The main feature is a shaft drive from the front axle operating the dropping mechanism, a brush or metal fixed wheel, in the hopper, while a lever attached to the handles could cut off the flow of seed when required. The rear wheel firms the soil over the seed and the rod extending at right angles has a sliding drag that marks out the line of the next row.*

clover, a hand-operated version, known as a shandy barrow, was often preferred. It was lightly constructed around a wheelbarrow frame supporting a seed box 11 feet (3.35 m) long with a hinged lid.

Seed was discharged from the hopper through a line of openings by a series of circular brushes rotating on a single shaft. This was set in the bottom of the box and connected by a bevel gearing to the axle of the barrow wheel. A pivoting lever on the frame acted as a clutch to throw the connecting pulley in and out of gear and prevent the barrow from distributing seed when being pushed from one field to the next. Swivelling copper plates, bearing one large hole and a series of smaller ones, were positioned over the discharging orifices and could be adjusted to regulate the amount of seed sown.

DRILLING

Small hand-operated seed drills, mounted on barrows, used similar arrangements for driving the dropping mechanism from the main axle. Many locally made types have survived, with one or two ground wheels, for sowing a single row of turnips, peas or beans. A band, twisted chain or shaft with bevel gearing took the drive back to the base of the seed box and a disengaging lever operated at either the axle or hopper end. Seed was discharged into the tinned chute beneath the hopper, usually by a rotating grooved cylinder or circular revolving brush. Alternatively, the seed hopper itself could be made in the shape of a small barrel and rotated to drop seed through a line of adjustable holes in the centre of its circumference and into the chute.

DIBBLING

Planting seed and small plants with a dibble, often fashioned from a shortened spade handle with a pointed end, is still common practice in the vegetable garden. But field dibbling of grain was quite widespread in the nineteenth century, particularly on lighter soils, and is reported to have originated in south Norfolk in the mid 1770s. The worker walked backwards across the field with a dibble in each hand, making two lines of conical impressions, 1 inch (25 mm) deep, in the soil at intervals of 3 inches (75 mm), with 4 inches (100 mm) between the rows. As the implement was withdrawn from the earth it was given a slight twist to ensure that the hole was smooth and clear to receive seed. Following behind came two droppers, often children, to drop between two and seven seeds into each hole.

Dibbling was a slow, labour-intensive process prone to a range of faults unless closely supervised. Achieving straight

Standard form of dibble with wrought iron shaft and bulb and wooden grip on the D-shaped handle. 32¼ inches (820 mm) long.

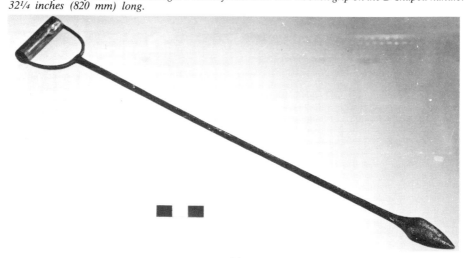

Women hoeing between the rows of plants in a field of peas, 1952.

rows was difficult and if the droppers were slack the wrong number of seeds might be placed in each hole or holes might be missed out altogether. Whether or not dibbling was worthwhile was a matter of debate, but its supporters argued that the even spacing and planting at regular depths not only saved seed but also gave higher yields to justify the heavier outlay on labour.

HOEING

The widespread adoption of the seed drill during the nineteenth century facilitated the use of horse hoes to clean the land between the rows of growing plants. Hand hoeing to remove weeds was still necessary, however, between the plants in each row and where seed was sown broadcast or in rows that were too closely bunched to admit horse-drawn implements. The hoe had a number of additional functions: for both loosening soil and drawing it up around the base of plants to stimulate growth; for singling rows of plants such as turnips, so that only a proportion were allowed to reach maturity; and for making shallow trenches before sowing seed in rows. In

consequence, an enormous range of steel hoes, partly based on regional patterns, was available from the Sheffield edge-tool manufacturers by the end of the nineteenth century.

Inwardly curving blades were advocated in the 1850s but there were many other varieties, some with blades perforated so that they did not collect soil or weeds together in heaps. Later catalogues concentrate on solid, flat-bladed hoes, which worked when pushed or pulled and were more suited to earthing up and thinning. Nevertheless, blades varied widely in size and shape according to whether they were required for scratching the surface of light soil or more deeply penetrating heavy land. Turnip singling called for a broad blade, perhaps 7 inches (180 mm) wide, that was pushed into the row to clear a gap of 7 inches on either side of each remaining plant. The Dutch hoe, used solely with a pushing action, was principally an implement of the garden, rather than the farm, but was convenient for some operations, as in the weeding of a young hedge, where only a very shallow disturbance of the soil was necessary.

ABOVE: *Hoeing turnips with broad-bladed swan-neck hoes in Perthshire, 1937.*

CENTRE: *A double-headed hoe enabled both sides of a row to be weeded simultaneously. It was heavy and cumbersome to use, however, and so is relatively uncommon. 13½ inches (340 mm) across.*

BOTTOM: *A weed hook from Suffolk, used for weeding in amongst the standing corn. It contains a socket for a long wooden handle. Total length of hook 20¾ inches (528 mm).*

Bird scaring in the early twentieth century. Protection of the young plants was often a task allocated to small boys and was their first introduction to working life on the land. The clappers for bird scaring consist of a central bat of chestnut, to which an oblong piece of deal is attached on either side with a leather thong. Total length 14 inches (356 mm).

BEST STEEL SPUDS & WEED HOOKS

"SKELTON" BRAND

ALL BRIGHT SOLID STRAPPED SLASHER SPUDS

12044

BLADES   2 inches Wide   4 feet Polished Knob Ash Handles   30 doz.

12045

BLADES   2 inches Wide   4 feet Polished Knob Ash Handles   30 doz.

STAMPED OPEN SOCKET WEED HOOKS

12050

Up to 2	2¼	2½ inches
5 3	5 9	6 3 doz.

12053

ABOVE: *From the 1915 catalogue of Skelton and Company, Sheffield.*

BELOW LEFT: *A dock lifter from Nottingham. The two prongs were thrust down either side of the stem using the foot rest for extra force. The tool could then be pivoted on the loop of wrought iron to lever the whole plant, including the root, out of the ground. 40 inches (1013 mm) long.*
BELOW RIGHT: *A twentieth-century sickle, manufactured by John Harrison and Son of Sheffield, with a serrated blade of cast steel. Blade 21¾ inches (550 mm) long.*

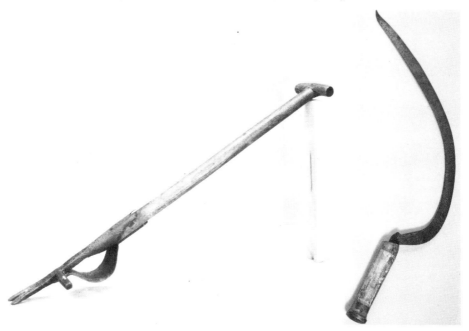

A large cradle attachment, reinforced with sacking, on a scythe being used to mow wheat.

HARVESTING

REAPING

The balanced sickle has been known for at least three thousand years and until the nineteenth century was the principal tool for cutting cereal crops in Britain. Its precise form was evolved over many centuries to sustain efficiency by minimising the accumulative effort of wielding it in the harvest field for many hours at a time. This led to the distinctive shape, with the point of balance located where the backward arch of the blade changed to a forward curve, and the use of materials, notably cast steel, that combined lightness with strength and durability. The blade did not require sharpening for it was edged with saw-teeth formed from a series of angled serratures on its underside.

Cutting was achieved with little more than a smooth circular action of the wrist that brought the edge of the curving blade into contact with the corn at the same time along all its length. For the sickle to be an effective implement, however, the reaper worked from a crouching position and had first to hold rigidly in his other hand the stalks that were to be cut. Each clutch of corn was then laid on a band of straw ready to be bound into a sheaf when enough had accumulated. Output was low with between only a quarter and a third of an acre (0.1 to 0.13 ha) being harvested by a reaper in a day. It was because the sickle could be worked by women as well as men that it survived into the twentieth century on very small holdings where the family constituted the primary labour force.

The smooth-edged sickle, or reaping hook, was of similar shape but the blade

LEFT: *Reaping with sickles, from William Hone's 'The Year Book of Daily Recreation and Information', 1878. This illustrates the stooping position of the reaper grasping a bunch of corn with his left hand while the cut is made with the serrated edge of the sickle blade.*
RIGHT: *When harvesting by hook, a smooth cutting action of the blade was often aided by the use of a crooked stick to lay the stalks away against the standing crop and then after each stroke to gather them into a bunch on the ground.*

was broader and the edge was honed sharp rather than serrated. This allowed the reaper to sweep through the crop using the momentum of the blade alone, without having to hold the stalks in individual bunches. However, work was interrupted more frequently to apply the sharpening stone to the underside of the blade and so maintain its edge.

BAGGING

The blade of the bagging hook was broader and heavier with a slightly more open curve. It was used with a pronounced swinging stroke, penetrating up to 1 yard (1 m) into the standing corn, in contrast to the careful and confined cut of the sickle. The bagger moved across the face of the crop, and at right angles to the direction of cut, pushing the stalks back

away from him with a hooked stick before each stroke. At the end of his line he walked backwards rolling the corn just cut under his foot until he reached his starting point with a bundle big enough to make half a sheaf. The procedure was then repeated.

In some of the southern counties of England bagging hooks replaced sickles for harvesting wheat during the middle of the nineteenth century. This was in response to a growing scarcity of labour for, at a cutting rate of over half an acre (0.2 ha) per day, the bagging hook was twice as fast as the sickle. It also cut closer to the ground, thereby providing more straw for use at the farmstead in the preparation of the all important livestock manure. Its disadvantages were that it left untidier fields of uneven stubble and

ABOVE: *Scything hay from 'Rural England' by L. S. Sequin, 1885. The four figures illustrate progressively how the scythe was swung around the mower's body. The handles can slide along the S-shaped sned and are fixed by wedges. Normally, the right-hand handle was positioned one blade's length above the base of the sned while the upper one was higher by a distance equal to the length of the mower's forearm.*

BELOW: *Two hooks from the Skelton catalogue of 1927. A cranked blade was a matter of personal preference but did allow the cutting edge to sweep more comfortably parallel to the ground.*

was too heavy a tool to be wielded satisfactorily by women. Greater speed left, in addition, more stray stalks and loose corn on the ground to be raked together by the pair of labourers who bound and stooked the sheaves produced by each bagger.

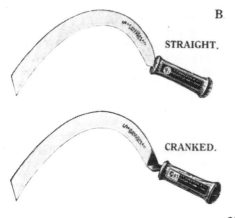

B

STRAIGHT.

CRANKED.

MOWING

The scythe was used continuously from the Roman period until recent times to mow grass for hay, but more intermittently in the corn harvest. Although it had previously sometimes been used on oats and barley, only in the nineteenth century did the scythe, with its faster work rate, challenge the supremacy of the sickle for harvesting wheat. From the 1850s onwards, however, many farmers who had converted to the scythe subsequently changed to the reaping machine so that by the end of the century only one fifth of Britain's corn crop was cut by hand.

The characteristic S-shaped shaft, or sned, of the scythe, enabling it to be swung around the mower's body, evolved during the twelfth century. Made usually out of willow, it became common in England, although the older and straighter form survived longer in northern Britain. For mowing corn, a loop of hazel or a light wooden frame was attached to

ABOVE: *Mowing corn with the scythe, from 'The Book of the Farm', by Henry Stephens, 1851, volume II. In this arrangement, three mowers are followed by three women gatherers to make the straw bands and collect the crop into sheaf-sized bundles. The sheaves are tied and built into stooks by the three bandsters while the tenth member of the team uses the stubble rake to collect together the stray stalks before tying them in a separate bundle. Three types of scythe are depicted, with the S-shaped sned on the left, the straight sned in the middle and the two-handled Aberdeenshire scythe on the right.*

LEFT: *A scythe strickle and horn from Brecon, Powys. The strickle is a tapering four-sided piece of oak fashioned to include a handle. Its surfaces were smeared with a mixture of grease and fine sand, a supply of which was kept in the horn. The result was an abrasive tool suitable for maintaining a keen edge on a scythe blade out in the field.*

the base of the sned. This device helped to lay the bunch of cut corn to one side in a neat pile, thereby easing the labour of the gatherers.

The scythe could mow an acre (0.4 ha) of wheat in a day and mowed closer to the ground to produce a greater quantity of straw than was possible with the sickle. Oat straw was cut with little effort so here the rate increased to 2 acres (0.8 ha) a day but the figure was slightly less with barley for the stalks took the edge off the scythe blade so that more time was spent on sharpening. Mowing was physically arduous and regarded as men's work although women and children often comprised the teams that followed behind to gather, bind and stook the sheaves as well as rake together the gleanings.

'Star' hand drag general purpose hay and stubble rake made in the 1880s by the Wexford Engineering Company of Ireland. Rake head 5 feet (1530 mm) wide.

RAKING

The stubble rake had a row of iron teeth set 4 inches (100 mm) apart and curving under at the lower end so that when drawn across the surface of the field it gathered only the loose corn stalks without disturbing the soil. The hay rake was an altogether lighter tool made entirely of wood. It survived long after the introduction of horse rakes in the early nineteenth century, especially on holdings possessing small or very hilly fields that could only be satisfactorily raked by hand.

Grass cut on the first day was tossed and turned by machine or hand fork, a process known as tedding, in the middle of the following day to expose it evenly to the drying action of the sun. Then it was raked together into windrows across the field and made into small round ricks or haycocks before nightfall. If the weather remained fair on the third day, the ricks were scattered again in the morning to complete the drying process in time for stacking during the afternoon. Under ideal conditions, the labour of haymaking was so arranged amongst the different operations that an equal amount was cut and stacked each day.

Hay rakes were used with a light skimming rather than a dragging action over the surface of the ground and their short ash teeth had rounded ends to prevent them digging in. The weakest part of the implement, where it was likely to break, was at the point where the ash shaft met the harder wooden head. One means of forming a stronger joint was to split the shaft towards one end to make it enter the head at two separate places. Alternatively, where a single joint was made, iron or wooden supporting braces were secured on either side across the diagonals between shaft and head.

STACKING AND ROPEMAKING

Haystacks were normally built to a regular rectangular shape of variable length but with the height proportional to the width of the base. Pitchforks were used for loading the carts or wagons in the field and again for feeding the hay up to the stack builder. The transit procedure had the beneficial effect of giving the hay a final airing to ensure that it had been correctly dried.

Before the stack was given a protective

23

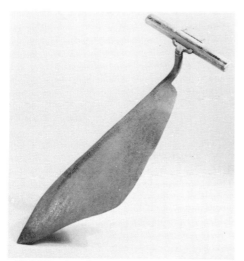

ABOVE: *A well worn Brades crown hay knife with cross handle. As the handle is set parallel to the blade, both hands can be used with maximum effect to produce the required chopping action down into the dense mass of hay. Handle 14 inches (355 mm) long.*

BELOW: *Stacking hay on the farm of Thomas Brown at Horley, Surrey, c 1910. Here hay from the wagon is being loaded first on to an intermediate platform, from where it is hoisted by fork up to the top of the stack.*

roofing of thatch the process of heat generation that occurs when new hay is compressed was allowed to run its course. When the hay was too damp, the fermenting temperature was liable to rise to a dangerous level and the subsidence to be so great and uneven as to threaten the stability of the stack. A thermometer, incorporated into a long iron tube and inserted into the centre of the stack, gave early warning of overheating so that preventive action could be taken to shore up the corners and admit a passage of cooling air. After settling, the stack was trimmed and squared and the eaves were given a slight overhang with an old scythe blade mounted on a long shaft.

When required, the hay was cut out in solid blocks, working downwards from one end of the stack so that the remainder retained its thatched covering. It was an arduous task, performed with the heavy broad-bladed hay knife. Used from a kneeling position, the knife was repeatedly thrust down into the hay to cut around the desired portion.

It was not uncommon for corn stacks to be built in the round from layers of sheaves arranged radially with their butt ends facing outwards. The stooks of corn remained in the field for longer, up to

Straw rope making, with an old bicycle wheel serving as the twister. Southern England, 1930s.

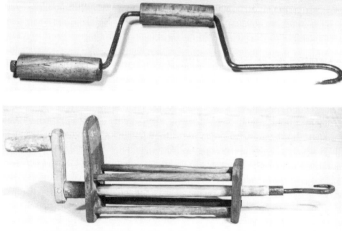

Straw rope twister, or throw crook, from Berkshire. Only experience could prevent the rope being twisted so tightly that it snapped or so loosely that it fell apart when used. 19 inches (482 mm) long.

Straw or hay rope twister from Durham, with cranked handle but straight wooden spindle enclosed in a cage. 24 inches (610 mm) long.

one week for wheat and two for barley and oats, to dry the straw and prevent decomposition in the stack. The thatching on the top of both hay and corn stacks consisted of overlapping mats of straw held down by a network of straw ropes, each about 30 feet (10 m) long.

Straw rope was much in demand at harvest time and was made on the farm with the throw-crook. A number of variants of this tool have survived but one common type had a cranked iron shaft with one or two cylindrical handles and a hooked end. The spinner placed a few lengths of straw in the hook and began twisting them by rotating the shaft. At the same time, the maker, seated beside a pile of loose straw, fed a continuous stream of stalks into the twisting length of new rope that was drawn out as the spinner walked slowly backwards away from him.

ABOVE LEFT: *Close-up of the head of a flail showing the jointing arrangement. The ash hand staff, left, has a swivelling cap made from a piece of ash steamed and bent to the required shape and bound on. Through the loop of the cap runs a leather thong linking the hand staff to the swingle, the head of which is lined with hide.*

ABOVE RIGHT: *The grid barley awner or hummeller was used with a stamping action upon threshed barley spread over a wooden floor. 37 inches (940 mm) high.*

LEFT: *The winnowing process began with the use of a wide-meshed riddle to separate out any stray pieces of straw and unthreshed ears of grain. This riddle, from Ashby de la Zouch, Leicestershire, has a screen of bamboo splits stretched across the beech frame, but a mesh of rushes or split willow was also common. 28 inches (710 mm) in diameter.*

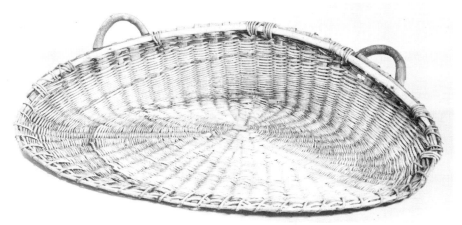

A winnowing basket of willow on an ash slot frame. The handles are of ash and the edging has been sewn with cane for greater strength. 53 inches (1350 mm) across by 12 inches (305 mm) at the highest point.

PROCESSING THE CROP

THRESHING

Threshing corn by hand — to separate the grain from the straw — was for so long a slow, laborious task that the first of many patented attempts to mechanise it was made in 1636. The working principles of a practical threshing machine, however, evolved during the second half of the eighteenth century and culminated in the patent of 1788 granted to the Scottish millwright Andrew Meikle. Over the following hundred years improvements in design and manufacture, together with the emergence of contractors hauling their portable threshers from farm to farm, brought mechanisation within the reach of all but the very smallest growers. For corn, therefore, the flail was gradually superseded during the second half of the nineteenth century, but for many years it was still retained, even on big farms, for threshing small quantities of vegetable and seed crops or producing a specimen sample of grain.

The procedure for threshing the corn crop by hand began with the transfer of one stack into the barn, where the sheaves were piled up in long lines or mows. An area lying between the double doors on either side of the barn was reserved for the threshing floor: the surface was often lined with close-fitting planks of oak or beech that lent a springiness to the action of the flail without breaking up under the constant pounding. The long ash staff of the flail, held in both hands, was first raised above the head and then swung forwards to bring the shorter beater rod, of ash or thorn, down upon the corn spread on the floor. As little more than seven bushels of wheat were the result of a day's hard work, the threshing of a farm's entire crop could occupy a large labour force over the winter period.

After barley had been threshed, it was hummelled to remove the short lengths of awn still attached to the grain. Hummelling tools either stamped or rolled upon barley spread on the floor to knock off the awns but hand-powered hummelling machines had been developed by the 1830s and subsequently adopted a variety of forms. In most, however, the awns were separated by blunt knives revolving on a spindle as the grain passed down an inclined cylinder or along a horizontal one.

WINNOWING AND DRESSING

The process of cleaning or winnowing began by riddling the corn through a wide

LEFT: *A nineteenth-century winnowing fan from Pewsey in Wiltshire. It consists of a series of spars, each equipped with a sail of sacking, mounted on a central spindle that can be rotated by hand with the aid of the heavy wooden flywheel. Fans of this type for maintaining a strong and constant passage of air first appeared in England in the late seventeenth century. The corn could be riddled in front of the fan or thrown from the shovel across its line of draught.*
RIGHT: *A nineteenth-century rotary barley awner from Essex. Roller 27 inches (685 mm) across.*

willow mesh to take out pieces of straw and any unthreshed ears that the flail had missed. Simple hand methods for removing chaff survived on some farms through the nineteenth century and all relied upon the ability of the wind to divide this lighter material from the heavier grain. When corn was tossed into the air from a shallow winnowing basket, for example, the grain fell back while the chaff was carried away by the draught of air flowing through the barn between the two sets of open doors.

A flat-bladed shovel, rather than the basket, was often preferred for throwing the corn against the direction of the wind. The chaff was blown away while at the same time the corn was graded into the best and heaviest sample that travelled furthest, the medium quality that fell short and the damaged or shrivelled grains, known as tailings, that landed closer still. Of these, the first was intended for market, the second often for home use and the third for livestock feed.

Market and seed corn underwent the supplementary stage of dressing to take out any remaining small stones, dust and weed seeds. It was passed firstly through a riddle with a wooden or wire mesh, 2⅜

inches (60 mm) square for wheat but larger for oats and barley, and then over a fine wire sieve or screen.

The hand-powered winnowing machine, incorporating a rotary fan of wooden blades inside a closed housing, originated in China over two thousand years ago. By means of trading links, it found its way to mainland Europe in the sixteenth century and from there to Britain by the early eighteenth century. Its use spread more quickly in the first half of the nineteenth century with improvements to the design for, whether corn was taken from the flail or threshing machine, two bushels could be winnowed in a few minutes rather than the hour that was required using more basic methods. If shaking riddles and screens were added to the hand-crank mechanism, grading and dressing could be carried out at the same time as the winnowing, although two or more passes through the machine were usually necessary to produce a clean sample for market. Steam-driven combined threshing and dressing machines were available from 1848 and gradually became established but the hand-operated dressers continued in use well into the twentieth century.

CHAFF CUTTING

When hay or straw was chopped into chaff, small pieces ranging from ¼ to ¾ inch (6-20 mm) in length, it was more easily and fully digested by livestock. The cutting box appeared around 1760 and over one hundred years later was still widely used, for it represented the simplest device available. Later in the eighteenth century, a sprung block or bar was added at the mouth of the trough. Raised and lowered by the operator through a foot pedal, it compressed and held the straw rigid before each downward stroke of the knife.

In the 1790s the practice developed of mounting one or two curved cutting blades on the spokes of a flywheel set at right angles to the trough. This not only increased the potential cutting rate but also allowed auxiliary operations to be driven through gearing from the rotating spindle at the centre of the wheel. An endless cloth belt running around a roller at either end of the trough, for example, moved the hay towards the cutting blades at a synchronised speed. Subsequently two spiked metal rollers, one on top of the other, were set at the mouth of the box to feed the hay in. Apart from more detailed refinements and the substitution of cast iron for wood in the body frame, the hand-operated chaff cutter changed little throughout the nineteenth century and was manufactured in large numbers by many British firms.

PROCESSING ROOTS AND OTHER FEEDS

Equipment for breaking up root crops, such as turnips, was necessary to aid their consumption by livestock. Sheep, in particular, suffered from damage to and even loss of their front teeth during their lives and so benefited from feeding on turnips in a prepared form. There was also a strong element of economy for when whole roots were supplied their insides were scooped out by the animals to leave a wasteful residue or hull. The turnip chopper was thrust down on to turnips lying on the ground and cut each into four pieces; if the action was

LEFT: *A chaff cutter of the late nineteenth century. The two curved cutter blades are mounted on the cast iron flywheel, which is turned by hand. Hay or straw is fed into the cutting position through two spiked rollers that are rotated by gearing from the flywheel shaft.*

RIGHT: *A hand-operated root slicer by F. Randell Ltd of North Walsham, Norfolk. The cutting was achieved by the combined action of two sets of cast steel knives, one mounted on the lever and one on the cast iron frame. Priced at £2 5s in the 1896 catalogue.*

LEFT: *Chopping turnips for sheep in the 1930s with a 'Gardner' pattern root cutter by Blackstone and Company of Stamford, Lincolnshire. This double action version cut wide slices for cattle when the barrel was rotated in one direction and narrow finger pieces for sheep and calves when turned in the opposite direction.*

RIGHT: *A turnip chopper with cruciform wrought iron blade capable of cutting each root into four pieces at one stroke. Total length 32¼ inches (820 mm).*

repeated, the separate pieces were similarly divided. This was strenuous work on a crop of any size and worthy of mechanisation. A device resembling the early box chaff cutter presented some improvement while another sliced turnips between two interlocking sets of cutting bars that were levered together.

All these early methods, cutting only one turnip at a time, survived but from the 1820s more elaborate and productive machines were being developed. Most incorporated rotating or oscillating blades that were driven from a hand crank and arranged vertically to one side or horizontally beneath a hopper. A second series of knives, placed behind the first, cut the resulting slices into finger-shaped pieces that both sheep and cattle preferred.

By mid century, one of the most highly recommended machines was that made by Gardener of Banbury. The turnips were loaded into a hopper, where they pressed down upon a revolving cylinder and were cut into fingers by two stepped arrangements of blades running across

the surface on either side. The revolving cylinder principle also applied to turnip pulpers except that rows of curved hooks, rather than blades, shredded the roots. These were then mixed with chaff in the cattle feed troughs.

During the 1820s improvements were made in other areas of feed preparation. They were thereafter refined and marketed in the middle of the century by the specialist agricultural engineering enterprises, such as Hornsby, Bentall and Crosskill, that were gaining national prominence. Mills to render oats, beans and peas more palatable by bruising them between two fluted rollers and crushers for breaking up linseed cake for cattle consumption were products of this period. They were all offered initially in the hand-operated form and continued to be so, but progressively new developmental work was concentrated on machinery with the strength and capacity to perform under the higher speeds made possible by horse, steam and, as the nineteenth century came to a close, internal combustion engine power.

PLACES TO VISIT

Intending visitors are advised to establish dates and hours of opening before making a special journey.

Abbot Hall Museum of Lakeland Life and Industry, Kendal, Cumbria LA9 5AL. Telephone: Kendal (0539) 22464.

Acton Scott Working Farm Museum, Wenlock Lodge, Acton Scott, Church Stretton, Shropshire. Telephone: Marshbrook (069 46) 306.

Alscott Farm Museum, Shebbear, Devon. Telephone: Shebbear (040 928) 206.

Ashley Countryside Collection, Wembworthy, Chulmleigh, Devon. Telephone: Ashreigney (076 93) 226.

Ashwell Village Museum, Swan Street, Ashwell, Baldock, Hertfordshire.

Barleylands Farm Museum, Barleylands Road, Billericay, Essex.

Beck Isle Museum of Rural Life, Beck Isle, Pickering, North Yorkshire YO18 8DU. Telephone: Pickering (0751) 73653.

Bickleigh Mill Farm, Bickleigh, Tiverton, Devon. Telephone: Bickleigh (088 45) 419.

Bicton Park Countryside Collection, East Budleigh, Budleigh Salterton, Devon. Telephone: Colaton Raleigh (0395) 68465.

Bodmin Farm Park, Fletchers Bridge, Bodmin, Cornwall. Telephone: Bodmin (0208) 2074.

Breamore Countryside Museum, Breamore House, Breamore, Fordingbridge, Hampshire. Telephone: Downton (0725) 22233.

Bygones at Holkham, Holkham, Wells-next-the-Sea, Norfolk. Telephone: Fakenham (0328) 710806.

Cambridge and County Folk Museum, 2/3 Castle Street, Cambridge CB3 0AQ. Telephone: Cambridge (0223) 355159.

Carron Row Farm Museum, Segensworth Road, Titchfield, Fareham, Hampshire PO18 5DZ. Telephone: Fareham (0329) 45102.

Castle Cary and District Museum, Castle Cary, Somerset

Clitheroe Castle Museum, Castle Hill, Clitheroe, Lancashire. Telephone: Clitheroe (0200) 24635.

Cogges Farm Museum, Church Lane, Cogges, Witney, Oxfordshire. Telephone: Witney (0993) 72602.

Cornish Museum, Lelant Model Park, Lelant, St Ives, Cornwall. Telephone: Hayle (0736) 752676.

Cotswold Countryside Collection, Northleach, Cheltenham, Gloucestershire GL54 3JH. Telephone: (summer) Cotswold (0451) 60715, (winter) Cirencester (0285) 5611.

Cotswold Folk and Agricultural Museum, c/o Andrews Cottage, Asthall Barrow, Burford, Oxfordshire. Telephone: Burford (099 382) 2178.

Country Life Museum, Sandy Bay, Exmouth, Devon. Telephone: Exmouth (0395) 274533.

Craven Museum, Town Hall, High Street, Skipton, North Yorkshire BD23 1AK. Telephone: Skipton (0756) 4079.

Dairy Land, Tresillian Barton, Summercourt, Newquay, Cornwall. Telephone: Mitchell (087 251) 246.

Dorset County Museum, High West Street, Dorchester, Dorset DT1 1XA. Telephone: Dorchester (0305) 62735.

Elvaston Farm Museum, The Working Estate, Elvaston Castle Country Park, Elvaston, Derby DE7 3EP. Telephone: Derby (0332) 73799.

Finch Foundry Trust and Sticklepath Museum of Rural Industry, Sticklepath, Okehampton, Devon. Telephone: Okehampton (0837) 84352.

Finkley Down Farm and Country Park, Andover, Hampshire SP11 6NF. Telephone: Andover (0264) 52195.

The Great Barn, Avebury, Marlborough, Wiltshire SN8 1RF. Telephone: Avebury (067 23) 555.

Guernsey Folk Museum, Saumarez Park, Catel, Guernsey. Telephone: Guernsey (0481) 55384.

Hampshire Farm Museum, Manor Farm, Upper Hamble Country Park, Botley, Hampshire. Telephone: Botley (048 92) 87055.

Highland Folk Museum, Duke Street, Kingussie, Inverness-shire PH21 1JG. Telephone: Kingussie (054 02) 307.

Hunday National Tractor and Farm Museum, Newton, Stocksfield, Northumberland. Telephone: Stocksfield (0661) 842553.

Ingram-Fowler Country Life Museum, Cricket St Thomas Wildlife Park, Chard, Somerset. Telephone: Ilchester (0935) 840103.

The Iron Mills. A. Morris & Sons (Dunsford) Ltd, The Iron Mills, Dunsford, Exeter, Devon. Telephone: Christow (0647) 52352.

Mary Arden's House, Wilmcote, Stratford-upon-Avon, Warwickshire. Telephone: Stratford-upon-Avon (0789) 3455.

Melton Carnegie Museum, Thorpe End, Melton Mowbray, Leicestershire. Telephone: Melton Mowbray (0664) 69946.

Model Farm Folk Collection, Wolvesnewton, Chepstow, Gwent. Telephone: Wolvesnewton (029 15) 231.

Museum of East Anglian Life, Abbots Hall, Stowmarket, Suffolk IP14 1DP. Telephone: Stowmarket (0449) 612229.

Museum of English Rural Life, The University, Whiteknights, Reading, Berkshire RG6 2AG. Telephone: Reading (0734) 875123 extension 475.

Museum of Lincolnshire Life, The Old Barracks, Burton Road, Lincoln LN1 3LY. Telephone: Lincoln (0522) 28448.

Norfolk Rural Life Museum, Beech House, Gressenhall, Dereham, Norfolk. Telephone: Dereham (0362) 860563.

Norris Museum, The Broadway, St Ives, Huntingdon,Cambridgeshire PE17 4BX. Telephone: St Ives, Cambridgeshire (0480) 65101.

North Cornwall Museum and Gallery, The Clease, Camelford, Cornwall PL32 9PL. Telephone: Camelford (0840) 212954.

North of England Open Air Museum, Beamish, Stanley, County Durham DH9 0RG. Telephone: Stanley (0207) 231811.

Old Kiln Agricultural Museum, Reeds Road, Tilford, Farnham, Surrey GU10 2DL. Telephone: Frensham (025 125) 2300.

Oxfordshire County Museum, Fletcher's House, Woodstock, Oxfordshire OX7 1SN. Telephone: Woodstock (0993) 811456.

Park Farm Museum, Milton Abbas, Dorset. Telephone: Milton Abbas (0258) 880216.

Pennine Farm Museum, Ripponden, Sowerby Bridge, West Yorkshire HX6 4DF. Telephone: Halifax (0422) 54823 or 52334.

Priest's House Museum, Wimborne Minster, Dorset. Telephone: Wimborne (0202) 882533.

Rutland County Museum, Catmos Street, Oakham, Rutland, Leicestershire. Telephone: Oakham (0572) 3654.

Ryedale Folk Museum, Hutton-le Hole, York YO6 6UA. Telephone: Lastingham (075 15) 367.

Scolton Manor Museum, Scolton, Spittal, Haverfordwest, Dyfed. Telephone: Clarbeston (043 782) 328.

Scottish Agricultural Museum, Ingliston, Newbridge, Midlothian. Telephone: 031-333 2674.

Shaftesbury Local History Museum, Gold Hill, Shaftesbury, Dorset. Telephone: Shaftesbury (0747) 2157.

Somerset Rural Life Museum, Abbey Farm, Chilkwell Street, Glastonbury, Somerset BA6 8DB. Telephone: Glastonbury (0458) 32903.

Staffordshire County Museum, Shugborough, Stafford ST17 0XB. Telephone: Little Haywood (0889) 881388.

Stewartry Museum, St Mary Street, Kirkcudbright, Telephone: Kirkcudbright (0557) 30797.

Swaledale Folk Museum, Reeth Green, Reeth, Richmond, North Yorkshire DL11 6RT. Telephone: Richmond (0748) 84373.

Temple Newsam Home Farm, Leeds, West Yorkshire LS15 0AD. Telephone: Leeds (0532) 645535.

Upminster Tithe Barn Agricultural and Folk Museum, Hall Lane, Upminster, Essex. Telephone: Romford (0708) 44297.

Upper Dales Folk Museum, Station Yard, Hawes, North Yorkshire. Telephone: Hawes (096 97) 494.

Weald and Downland Open Air Museum, Singleton, Chichester, West Sussex. Telephone: Singleton (024 363) 348.

Welsh Folk Museum, St Fagans, Cardiff, South Glamorgan CF5 6XB. Telephone: Cardiff (0222) 569441.

Weston Park, Shifnal, Shropshire. Telephone: Weston-under-Lizard (095 276) 207.

West Yorkshire Folk Museum, Shibden Hall, Halifax, West Yorkshire HX6 6XG. Telephone: Halifax (0422) 52246.

Wimpole Home Farm, Arrington, Royston, Hertfordshire SG8 0BW. Telephone: Cambridge (0223) 207257.

Yorkshire Museum of Farming, Murtonpark, Murton, York YO1 3UF. Telephone: York (0904) 489731.